Education
249

快乐五丑

The Ugly (but Happy) Five

Gunter Pauli

[比] 冈特·鲍利 著

[哥伦] 凯瑟琳娜·巴赫 绘

何家振 译

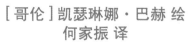

上海远东出版社

丛书编委会

主　任：贾　峰

副主任：何家振　闫世东　郑立明

委　员：李原原　祝真旭　牛玲娟　梁雅丽　任泽林

　　　　王　岢　陈　卫　郑循如　吴建民　彭　勇

　　　　王梦雨　戴　虹　靳增江　孟　蝶　崔晓晓

特别感谢以下热心人士对童书工作的支持：

匡志强　方　芳　宋小华　解　东　厉　云　李　婧

刘　丹　熊彩虹　罗淑怡　旷　婉　杨　荣　刘学振

何圣霖　王必斗　潘林平　熊志强　廖清州　谭燕宁

王　征　白　纯　张林霞　寿颖慧　罗　佳　傅　俊

胡海朋　白永喆　韦小宏　李　杰　欧　亮

目录

Contents

一头疣猪正快乐地在泥里打滚，笑得合不拢嘴。一只斑鬣狗走到他跟前说：

"你的脸多么丑啊，不管你笑不笑……"

A warthog is enjoying wallowing in the mud, and is smiling from ear to ear. A spotted hyena walks up to him and says: "What an ugly face you have, smile or no smile..."

一头疣猪正快乐地……

A warthog is enjoying ...

你没有看过自己在水中的倒影吗?

Have you ever looked at your reflection ...

"你好意思说我？" 疣猪回答道。"你没有看过自己在水中的倒影，也没听过自己的笑声吗？你夜里咯咯嗝嗝的叫声瘆得所有人脊背发颤。"

"我可不在乎，因为灌木丛中有那么多其他被认为是丑陋或者吓人的动物。就拿那边的角马来说吧，他可从来都不是谁的宠儿。"

"Look who's talking!" the warthog replies. "Have you ever looked at your reflection in the water, or heard yourself laugh? Your cackling and whooping send shivers down everyone's spine at night."

"See if I care, with so many other animals in the bush considered ugly or scary. Take that wildebeest over there, for instance. He's never been anyone's favourite."

"嗯，我们不可能都像花豹那样好看，也没有鱼鹰那样美妙的声音。在生活中一个简单的事实是，不是每个人都能让所有人觉得好看或者好听。"

"是的。就像我们的邻居，那边那只秃鹫。他肯定不是那些来观赏狮子、花豹、犀牛、水牛和大象的游客的宠儿。"

"Well, we cannot all have a leopard's good looks, or a fish eagle's interesting call, you know. It is a simple fact of life that not everyone looks or sounds good to everyone else."
"True. Like our neighbour, that vulture over there. He is certainly not a favourite with all the tourists who come to admire our lion, leopard, rhino, buffalo and elephant."

……那边那只秃鹫……

... that vulture over there ...

但是那些小生物呢？

But what about the small ones?

"嘿，你正好说出了最令人赞叹的五大动物。但是那些小生物呢？还有那些丑陋却快乐的动物呢？更不用说植物和菌类了。它们在大自然中都扮演着独特的角色。"

"你说得对，疣猪。如果没有他们，我们会怎么样？哪怕是最小的生物对这个大草原上的生命来说也是必不可少的。"

"Hey, you've just listed the Big Five, those that people find most impressive. But what about the small ones? Or the ugly, but happy, ones? Not to mention all the plants and fungi. They all have a unique role to play in Nature."

"You are right, Warthog. Where would we be without them all? Even the smallest ones are essential to life here on the savannah."

"你是说微生物和真菌吗？" 疣猪问。

"不，我说的是蚁狮、犀牛甲虫、牛文鸟、豹纹陆龟和象鼩这样的小动物。都很小，但又很独特……"

"是的！谁说一定得个头大才能对地球上的生命作出大的贡献？"

"Are you now talking about microbes and fungi?" Warthog asks.

"No, I am talking about creatures like the ant lion, the rhino beetle, the buffalo weaver, the leopard tortoise, and the elephant shrew. All quite small, yet so unique…"

"Yes！ Who said you have to be big to make a big contribution to life on Earth?"

谁说一定得个头大……

Who said you have to be big ...

……我最喜欢犀牛甲虫……

... the rhino beetle is my favourite ...

"刚说的'五小'，我最喜欢犀牛甲虫。小伙子们，姑娘们，这里有一只值得一看的甲虫。我很想知道它为什么长得这么特殊，为什么这么有劲儿……"

"我最喜欢的是牛文鸟。它们吃很多昆虫，而且建造巨大的公共巢穴。"

"说到鸟类，我很崇拜非洲秃鹳。"

"Of the Little Five I've just listed, the rhino beetle is my favourite. Boys and girls, now here is a beetle worth looking out for. I do wonder where it got its exceptional look, and its strength…"

"And my favourite is the buffalo weaver. They eat a lot of insects, and build huge communal nests."

"Well, when it comes to birds, I do adore the marabou stork."

"崇拜？怎么会有人'崇拜'这种丑陋的鸟？它吃腐肉，通过往自己腿上排泄来降温。恶心！"

"听着，疣猪，情人眼里出西施。不管你喜不喜欢它们的长相或习性，所有物种在生态系统中的重要作用都应该受赞美。"

"当然，你又说对了。"疣猪承认。"的确，我们不应该根据外表或者某些特质评判别人。"

"Adore? How can anyone 'adore' such an ugly bird, one that eats carrion and pees and poops on its own legs to keep cool? Yuck!"

"Look Warthog, beauty is in the eye of the beholder. And whether you like their looks or behaviour or not, all should be celebrated for their important role in the web of life."

"You are, of course, right once again," Warthog admits. "It is true that we should not judge others on their appearance or traits."

... beauty is in the eye of the beholder ...

......把我们叫作"五丑"......

... call ourselves the Ugly Five ...

"疣猪，为什么不让角马、秃鹫和秃鹳也加入我们，把我们叫作'五丑'呢？我们可能很丑，但我们很高兴能成为地球生态系统的一部分……"

"喂，为什么会有人想加入丑八怪名单呢？"

"这是一个错误的问题！为什么有的动物就应该被看不起？仅仅因为别人对他的外表的看法吗？还是因为他们不理解这个动物的行为？我们应该与这种不公正作斗争。"

"Warthog, why don't we include Wildebeest, Vulture, and Marabou Stork in our group, and call ourselves the Ugly Five? We may be ugly, but we are just so happy to be part of the web of life…""

"Now why would anyone want to be on a list of the ugly?"

"That is the wrong question! Why should anyone be looked down upon, just because of others' opinions about his looks? Or because they do not understand his behaviour? We should be fighting such injustice."

"好啦，鬣狗，我已经准备好和你并肩作战了。"

"好！让我们为大家的权利站出来，为了让我们不被武断地评判。难道我们没有权利做我们自己，证明自己吗？"

"确实，我们每个人都应该这样做！"疣猪表示赞同。

"所以，让我们一起赞美一下'五小'和'五大'，还有我们快乐的'五丑'。我向你保证，人们会爱我们所有人的！"

……这仅仅是开始！……

"Well, I am right here, Hyena. Ready to fight at your side."

"Good! Let's stand up for everyone's right not to be judged. Don't we all have the right to be ourselves, and prove ourselves?"

"We do indeed. Every one of us!" Warthog agrees.

"So, let's then celebrate the Little Five and the Big Five, along with us, the Ugly (but Happy) Five. I guarantee you, people will love us all!"

... AND IT HAS ONLY JUST BEGUN!...

······ 这仅仅是开始！······

... AND IT HAS ONLY JUST BEGUN! ...

Did You Know?

你知道吗？

Tourists travel to Africa to go on safari and see the Big Five, but neglect to pay attention to the many other species of wildlife that may, at first glance, appear less appealing. It is time to celebrate more than only the big ones, and encourage discovery of other creatures.

游客去非洲旅行是为了观赏五大野生动物，却忽视了许多其他乍一看可能不那么吸引人的野生动物。不应该仅仅赞美大型生物，也要鼓励人们去发现其他生物。

The call of the spotted hyena can be heard 12 kilometres away. Hyenas live in clans that can have as few as three or up to as many as 80 members. They have a strict hierarchy, with the males at the bottom of the social strata.

斑鬣狗的叫声在 12 千米外都能听到。鬣狗生活在 3 至 80 个成员的群落中。它们有严格的等级制度，雄性处于底层。

The marabou stork has a 4 metre wingspan, and can live for up to 25 years in the wild. Its grey legs appear white, as it squirts excrement onto its legs. It is the evaporation of moisture, however, that has a cooling effect.

非洲秃鹳的翼展为 4 米，在野外可以存活 25 年。它灰色的腿看起来是白色的，因为它把排泄物喷到腿上。正是排泄物中水分的蒸发产生了冷却效果。

The wildebeest is one of the prey animals favoured by lion. Wildebeest are able to reach a running speed of 80 km/h, to outrun predators. The beginning of the breeding season coincides with the first full moon after the rainy season, suggesting a lunar influence.

角马是狮子喜爱的猎物之一。角马能够以每小时 80 千米的奔跑速度逃脱捕食者的追捕。角马的繁殖季节始于雨季后的第 1 个满月，这体现了月亮的影响。

The rhinoceros beetle is the strongest animal on the planet, being able to lift and transport objects that are 850 times heavier than itself. If applied to a human, it would mean being able to lift and carry a weight of 65 tons, using muscle power only.

犀牛甲虫是地球上最强壮的动物，能够举起和运输比自己重 850 倍的物体。如果以人类体重换算，相当于仅用肌力就能举起和搬运 65 吨的重量。

Elephant shrews can jump as high as one metre. They are unique in that females menstruate. Adult shrews leave scent trails for their offspring to find food.

象鼩能跳 1 米高。它们的独特之处在于雌性象鼩会来月经。成年象鼩会留下气味痕迹，让它们的孩子找到食物。

蚁狮通过挖一个螺旋形的小坑，创造了动物界最简单、最有效的陷阱之一。这种陷阱利用精确的静止角，能将猎物困在雪崩般的沙粒中。蚁狮的猎物可达到其自身的 2 倍大。

The ant lion creates one of the simplest and most efficient traps in the animal kingdom by digging a small pit in a spiral shape, with a precise angle of repose so that their prey is caught in an avalanche of sand grains. Their prey can be twice the ant lion's size.

豹纹陆龟主要是圈养繁育的，是世界上交易量最大的动物之一。龟壳上的斑点很像豹皮，会随着时间的推移而褪色。野生豹纹陆龟吃鬣狗的粪便以获取钙质。

The leopard tortoise is mainly bred in captivity, and is one of the world's most traded animals. The spots on the shell, resembling leopard skin, fade over time. Tortoises in the wild eat hyena faeces for its calcium content.

Is it acceptable to dislike someone just because of appearance?

仅仅因为外表就讨厌某个人，这是可以接受的吗？

What is required to be considered cute?

可爱的必备条件是什么？

Would you stand up against the injustice of mistreatment based on appearance?

你会反对因为外表不好看就被虐待的不公正现象吗？

Which is your favourite group: the Big Five, the Little Five or the Ugly Five?

你最喜欢的组合是"五大""五小"，还是"五丑"？

Do It Yourself!

自己动手！

How much do we know about the diversity of African wildlife? Design three posters, showing pictures of the Big Five, the Little Five and the Ugly (but Happy) Five. Now, do a survey by asking your friends and family members to name each of the animals. Provide some interesting information on each of the Small Five and the Ugly (but Happy) Five. Now, ask every person to name their favourite small animal, as well as their favourite 'ugly' animal. Ask them to also provide a reason for their choice. Now share your findings.

我们对非洲野生动物多样性了解多少？设计 3 幅海报，展示"五大""五小"和"（快乐的）五丑"的图片。现在，做一个调查，让你的朋友或家人说出每种动物的名字。提供一些关于"五小"和"（快乐的）五丑"的有趣信息。然后，请每个人说出他们最喜欢的小动物和他们最喜欢的丑动物。让他们为自己的选择说出一个理由。最后，分享你的调查结果。

学科知识
Academic Knowledge

生物学	五界分类系统将所有现存物种划分为：动物界、植物界、真菌界、原生生物界和原核生物界；界之下有多个门，如脊索动物门、节肢动物门、软体动物门和棘皮动物门；门之下有多个纲，如脊椎动物亚门下有哺乳纲、硬骨鱼纲、软骨鱼纲、鸟纲、两栖纲和爬行纲等；纲之下有多个目，如哺乳纲下有食肉目、灵长目、啮齿目和偶蹄目等；目之下有多个科，如食肉目下有猫科、犬科、熊科和鼬科等；科之下有多个属，如猫科下有猫属、豹属和猎豹属等；属之下有多个种；生命层次结构：细胞、组织、器官、器官系统、有机体；动物进食、消化和排泄。
化　学	动物呼吸中的生化反应将碳水化合物分解为二氧化碳和水，再转化成为细胞提供能量的三磷酸腺苷；动物体内的有机分子包括碳水化合物、脂类、蛋白质和核酸；碳水化合物是动物的主要能量来源。
物　理	动物如何处理热（狗喘气）、力（小丑虾以极大的加速度"出拳"）、液体（池龟水上行走）、声音（大象通过脚上肉垫接收声波）、磁场（红海龟利用地球磁场）、光（射水鱼使用光折射捕食昆虫）。
工程学	野生动物与人类栖息地工程解决人类与野生动物的冲突。
经济学	旅游业是非洲经济的引擎，例如在保护区观看五大野生动物；经济层面的生产力和韧性取决于以生物多样性为核心的生命维持系统；花在外表上的钱和被人喜欢的需要。
伦理学	只根据外表来判断一个人；如果一个人在社会中的价值，由该人能够并且已经做出的不可缺少的贡献来决定，那么人的外表又会有多大重要性；反对不公正的需要；学习自然知识或向自然学习。
历　史	在18世纪，林奈记述了4 400种动物和7 700种植物；一个世纪以前，五大动物是猎人的目标。
地　理	将生物分为五类的做法始于非洲；"五大""五小""五丑"的首选栖息地是大草原。
数　学	美国政府计算如何为了保护一些动植物而允许另一些动植物灭绝，并为此建立数学模型，其假设前提是不可能拯救所有濒临灭绝的物种。
生活方式	外貌在现代社会的重要性。
社会学	人们面对大自然会感到恐惧，因此需要探索，以达到新的舒适水平。
心理学	乐观、外向性格的魅力；婴儿的魅力（小而可爱）；"大五人格理论"的5个维度：经验开放性、尽责性、内-外向性、宜人性以及情绪稳定性；面对未知物种时的情感信息处理。
系统论	生态系统需要所有的物种，而不仅仅是最大的或最吸引人的物种。

情感智慧
Emotional Intelligence

鬣　狗

鬣狗以人身攻击的方式开始了对话。当他的侮辱受到回击时，他没有受别人对他的看法影响。鬣狗指出其他动物也因为外表而不受欢迎。他展示了对生物多样性和没有受到太多关注的物种的广博知识。他分享了自己对犀牛甲虫的喜爱，并鼓励其他人也来关注这种甲虫。他说明了自己对秃鹳的偏好，当秃鹳的坏习惯被提及时，他趁机就随意评判他人这一问题展开讨论。他成功地说服疣猪和他一起为正义而战。他相信，人们将学会赞美各种形式的生命，而不论其外表和习惯。

疣　猪

当疣猪因其外表受到侮辱时，他予以回击，并回敬说鬣狗叫声很难听。疣猪欣然承认，并不是每个人的外貌或声音都会让其他人喜欢。然后，疣猪迅速转换话题，让人们关注自然界中那些更小、更丑陋的生物，以及它们所扮演的独特角色。在鬣狗分享他最喜欢的动物之后，疣猪也说出他最喜欢的小动物。当鬣狗说出他最喜欢的鸟时，疣猪无法掩饰他对秃鹳的厌恶。尽管如此，他还是同意所有生物都扮演着自己的角色，因此不该以貌取人。疣猪已经准备好以更积极的姿态去对抗不公正。

艺术
The Arts

我们都见过五大动物的照片。现在让我们为"五小"和"（快乐的）五丑"分别创作一幅拼贴画，以这种方式来赞美它们。在互联网上搜索这些生物的图片，然后把每一组动物的图片编辑成有趣的组图。为了给你的拼贴画增添趣味，可以用能够体现每一种生物某些可爱的、有趣的和不寻常的特征和习性的图片。

思维拓展
Systems: Making the Connections

由于历史上对野生动物的大量捕杀，当前的气候变化，以及栖息地的丧失，许多物种已经灭绝。很多现存的物种也有灭绝的危险，需要采取措施保护它们。然而，只专注于少数大型和危险的动物物种，导致许多其他需要保护的物种受到的关注较少。健康的生态系统依赖于生物多样性。随着濒临灭绝物种数量迅速增加，一种逻辑正在显现：拯救所有极度濒危的动植物物种是不可能的，只能使用有限的资金去拯救某些特定物种。这是一个艰难的抉择，因为构成生命之网的每一个物种都需要同等的生存机会。虽然五大动物，或者最可爱的熊猫宝宝可能最受关注，但其他物种一旦灭绝，也可能严重影响整个生态系统。不仅需要紧急保护所有已知物种，还需要进行更多研究，以发现其他物种，研究它们与已知物种的相互作用，了解它们是如何为生态系统作贡献的。在这个探索的过程中，我们不能被短期需求所左右。我们需要考虑包括我们自己在内的所有物种的长期生存。探索我们不知道的物种是一项鼓舞人心的工作。我们甚至还没有开始认识到它们所提供的有关我们生存的有用信息是多么宽广和深奥。我们应该敬畏错综复杂的生态系统，尽一切努力确保所有物种都能继续沿着它们的进化道路前进，增强适应能力，并确保所有人过上满意的生活。所有生物都值得我们尊重和敬畏，而不仅仅是最可爱的或最危险的物种。引起人们对"五小"和"快乐的五丑"的关注可能是简单的第一步，但这将最终引发人们对所有形式生命的兴趣，包括植物和真菌。我们的生存依赖于尽可能多地发现与我们共同生活在这个星球上的所有生物的奇妙之处。

动手能力
Capacity to Implement

我们已经看了"五大""五小""五丑"，但还有更多有趣的生物。根据生物的特点选择并列出相应的生物，做一个展示生物多样性和促进保护所有物种的宣传活动计划。例如，你也可以列出"五稀""五快""五高""五趣"或"五幸"。每个类别列出5种生物，争取一共列出100个。这可不是件容易的事，因此，一定要寻求团队伙伴的帮助。

故事灵感来自
This Fable Is Inspired by

朱莉娅·唐纳森
Julia Donaldson

朱莉娅·唐纳森出生在伦敦。在她5岁的时候，她从父亲那里收到了一本《一千首诗》，此后，诗歌在她的早年生活中占据了重要地位。朱莉娅在布里斯托尔大学学习戏剧和法语。加入布里斯托尔街头剧场，对她与孩子们的互动产生了持久的影响。她创作了儿童音乐剧《格仑特国王的蛋糕》和《海盗在码头》。1977年，她毕业于布莱顿教育学院，获得了教育学硕士学位，之后成为了一名教师。2011年，布里斯托尔大学授予朱莉娅荣誉博士学位，2012年，格拉斯哥大学也授予朱莉娅荣誉博士学位。在她的畅销儿童故事《咕噜牛》中，朱莉娅从一个中国故事中获得了灵感。这个故事讲的是一个小女孩为了避免被老虎吃掉，声称自己是可怕的丛林女王，并邀请老虎走在她身后。2017年，朱莉娅写了《五只丑八怪》一书，启发我们写这个故事。

图书在版编目（CIP）数据

冈特生态童书.第七辑:全36册:汉英对照 /
（比）冈特·鲍利著;（哥伦）凯瑟琳娜·巴赫绘;
何家振等译.—上海:上海远东出版社,2020
ISBN 978-7-5476-1671-0

Ⅰ.①冈… Ⅱ.①冈… ②凯… ③何… Ⅲ.①生态
环境－环境保护－儿童读物—汉英 Ⅳ.①X171.1-49

中国版本图书馆CIP数据核字（2020）第236911号

策　　划　张　蓉

责任编辑　祁东城

封面设计　魏　来　李　廉

冈特生态童书

快乐五丑

〔比〕冈特·鲍利　著
〔哥伦〕凯瑟琳娜·巴赫　绘

何家振　译

记得要和身边的小朋友分享环保知识哦！
八喜冰淇淋祝你成为环保小使者！